EXPERIMENTS IN THE DETERMINATION OF MECHANICAL BEHAVIOR OF ENGINEERING MATERIALS

Seventh Edition

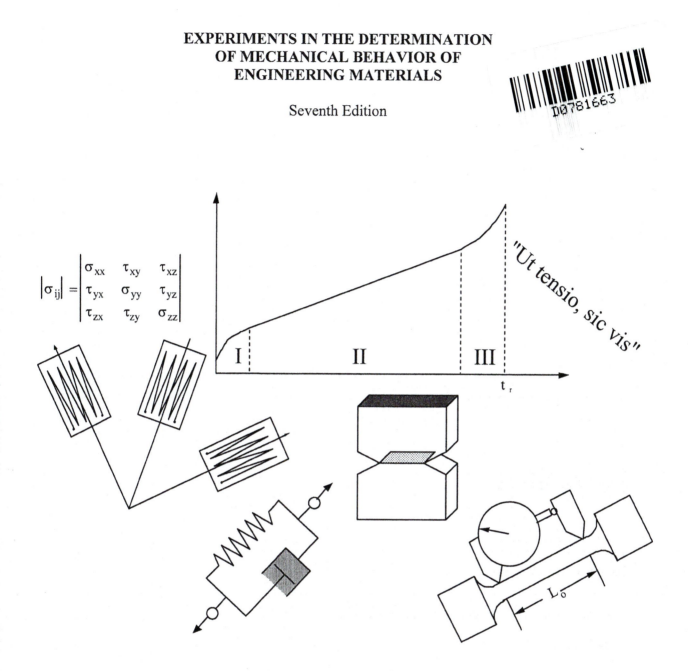

$$\left|\sigma_{ij}\right| = \begin{vmatrix} \sigma_{xx} & \tau_{xy} & \tau_{xz} \\ \tau_{yx} & \sigma_{yy} & \tau_{yz} \\ \tau_{zx} & \tau_{zy} & \sigma_{zz} \end{vmatrix}$$

"Ut tensio, sic vis"

Richard A. Queeney
Albert E. Segall
The Pennsylvania State University

Kendall Hunt
publishing company

Albert E. Segall's photos used on the cover

www.kendallhunt.com
Send all inquiries to:
4050 Westmark Drive
Dubuque, IA 52004-1840

Printed in the United States of America
10 9 8 7 6 5 4 3 2

DEDICATION

This lab book is dedicated to the memory of Richard A. Queeney, who has always been, and always will be its "heart and soul," regardless of how many changes are made down the road. Rick was a good friend who taught me as much about engineering as he did about life and unfortunately death. His great sense of humor and an innate ability to see the ironic (and yes, moronic) things in life, even when very ill, will be dearly missed. My only hope is that this book does Rick justice by continuing to be a good guide for engineering students, while still capturing his wit, knowledge, and love for the profession and all it entails!

ACKNOWLEDGEMENTS

Few things are possible without a cast of thousands, and this lab book is no exception. I would therefore like to acknowledge and thank the following who have helped me in so many ways that space and time will not allow a full rendering.

I would first like to thank Donna Queeney for her guidance and support that was a crucial mix of moral, legal, and financial.

I would also like to thanks Judith Todd of Engineering Science and Mechanics for her patience and assistance on so many levels I do not know where to begin.

Thanks also go to my wife Beth Segall, daughter Sarah Segall, and son Jake Segall who helped with typing, editing, and who all survived countless blank stares over the past few years.

Similar kudos go to Liliya Ventsel for her tireless help with teaching the labs and many great suggestions to improve them.

Last, but certainly not least, I would like to thank Lucas Passmore for creating the new figures used in this version and for technically editing the manuscript with many great suggestions.

CONTENTS

PREFACE

The testing of materials to determine their load bearing capabilities dates back to at least the time of Archimedes, who was mainly interested in the strength of wooden beams in the form of ships' keels. Such interests emerged as the shipbuilders of ancient Greece were embarked on a program of scaling up their rowing ships to more fun and profitable cargo and warships. Although the Hellenes were not privy to our current understanding of the concepts of stress and strain, they were certainly successful enough in building their triremes, defeating their enemies from the east, and evolving ways of interpreting nature. If you think about it, in many ways the Greeks have ensured that we would have the opportunity to exercise our own intellects in creating many truly useful things that people really need like the Salad Shooter or Pet Rocks to name a few.

Renowned engineers and scientists such as Leonardo da Vinci and Galileo reported testing materials in a way that we would immediately recognize as modern and more importantly, design oriented; Galileo actually tested iron wires in tension around 1500, while Leonardo used wooden cantilever beams a few decades later. However, it was not until the latter half of the 1800's that our present concepts of stress and strain became fully evolved (at least, we hope it is), thereafter defining the *what* and *how* of the mechanical testing of engineering materials.

One might then ask: to what mechanical tests, then, should load-bearing materials be subjected to? For this particular course, the answer clearly is only those tests whose quantitative results have direct application in engineering design. It may seem strange that testing is carried out for other reasons, but historical precedent, economics, and/or plain old (and unfortunately, never ending) human inertia often are the justification for these practices. For example, indentation hardness methods such as the Rockwell or Vickers, are quick, inexpensive, and somewhat nondestructive means for gauging some tensile parameters, particularly for comparative purposes. However, due to their analytical complexity that is often masked by physical simplicity, these tests are only indirectly utilized in design.

Given these concerns and dilemmas, the wonderful laboratory experience contained herein is not designed to expose the student to a few *properties* of a material, as these are defined as "characteristic traits of" or "qualities that serve to define" a material. Instead, we shall measure *responses* of a material to relevant service conditions so that they can indeed, be used as part of the design process. A simple example might be to answer, in a quantitative fashion, the question of how strong is a given material. The answer will depend on, in part:

> a. The geometry of the test specimen
> b. The stress or strain to be imposed on the specimen
> c. The rate at which it is loaded or deformed
> d. The temperature of testing and service
> e. The presence of aggressive chemical species
> f. The composition and fabrication history of the material and so on …

Hopefully, the student will extract from the laboratory experience an understanding of the role of mechanical testing in the design process, as well as an appreciation of the richness of the topic and how it impacts your daily lives. You may actually appreciate this knowledge, especially every time you dish out big "Bucks" to fix a materials related problem of your car or house!

REPORT WRITING

When asked as to what they expect an educated engineer to bring to the workplace, our "product" consumers (your eventual employers) inevitably mentions *competence in communication* as a trait of highest importance. You may know this important skill simply as the ability to write! Writing has always been, and probably always will continue to be the most crucial communication skill, no doubt accounting for its appearance at many points in your education. Fortunately, it is a skill that can be learned and improved through practice and exercise. Otherwise, the countervailing influence of such "high" culture media such as MTV, Dancing with the Stars, or even "Texting," would doom many otherwise promising young engineers to a lifetime of sales in mall stores, or even worse, a job where you get to ask the eternal question: do you want fries with that?

It may be surprising to many, but effective writing consists as much in knowing *what* to write as in knowing *how* to write it. For communicating technical ideas in the context of a project, there exist common formats whose mastery makes the entire report writing process relatively straightforward and low stress (all puns intended) in execution. While certainly not all inclusive, the format guide given below is an acceptable blueprint for reporting the type of technical data and outcomes typical for laboratory assignments.

REPORT FORMAT

Most reports will ultimately be judged complete if it contains all, or at least most of the sections listed below. In practice, however, readers will often examine only those sections of greatest relevance to them. For instance, most bosses and managers are notorious for quickly (and quietly) skipping to the conclusions, only reading the rest of the report if they have questions or need additional information. While this may lead the current reader to think that all that is required is the conclusions, the boss or manager will quickly chastise you (you hope that is all he/she does) if the rest of the report is not there. Hence, you should ignore human nature (including your own desire to do as little as possible) and make sure that the report contains some semblance of the following:

1. IDENTIFICATION

This is the easy part, kind of like making sure that your name is on an exam so you get credit. Identification is usually achieved through the use of a title page. In addition to the obvious (the title), this page should identify the date the experiment was performed, the name of the student(s), and the esteemed instructor since everyone likes getting credit.

2. OBJECTIVES

The objective is a statement of purpose, written to direct the reader to the reason for conducting the experiment in the first place. For example, Ben Franklin performed his kite flying experiments to determine whether lightning was a form of electricity. No such road to fame for you, as you might be conducting less dramatic experimental stress measurements to determine how stress concentration factors for notches depend on the notch depth and the notch tip acuity.

3. PROCEDURE

The procedure section should tell the reader what you physically did, and how you did it as succinctly as possible. Fortunately for us, western science and technology has this basic accountability built-in; someone with similar technical prowess must be able to duplicate your results to be accepted! Hence, your report must adequately describe how you actually did the test, what devices were employed and how were they used, what measurements did you make, and last but not least, what material did you characterize? Any given experiment may have a different list, but the basic idea remains the same.

4. RESULTS

Here, all that is to be communicated is just exactly what the section title states, *RESULTS*, pure and unadulterated in tabular and/or graphical form with ample description. Simply put, the numerical values of your experimental measurements are the results without data refinement. A common exception to this rule may be allowed if measurements are made in the U.S. Customary System, as they should be converted to the S.I. System and co-listed. Useful conversion factors are given below.

$$1 \text{ pound} = 4.448 \text{ Newtons}$$

$$1 \text{ inch} = 2.54 \text{ centimeters}$$

$$1 \text{ psi} = 6,895 \text{ Pascals}$$

$$1 \text{ ksi} = 6.895 \text{ MPa}$$

$$10^6 \text{ psi} = 6.895 \text{ GPa}$$

5. ANALYSIS OF RESULTS

The analysis section should contain all calculations performed on the as-measured data, such as converting load/deformation data to stresses and strains. A sample calculation should always be included, both for quick verification and as a standard professional practice. The sample calculation must be identified-e.g., "Flexural Stress"- and the units of all parts of the calculation, as well as the result, given. Remember that the number of significant digits in a calculated answer cannot exceed the number of significant digits in the quantities used in the calculation. Repetitive calculations, such as the conversion of load data to stress need only be demonstrated (fully) once. As before, the completely analyzed results can be given in tabular and/or graphical form. Remember, it has been said that a picture is worth a thousand words, so graphical presentations are often a very powerful way of conveying results and pointing out potential implications.

6. DISCUSSION OF RESULTS

As an engineer, you will be expected to draw and prove inferences to your system's response in a way that is generally useful. For example, if you had determined that the stresses at the tips of notches depend on the notch tip acuity, what quantitative generalizations can you make about notch severity and local stress? If you have measured a series of strength parameters for a metallic alloy, how do your values compare to the conventionally accepted measures from the literature? To keep the discussion focused, a list of pertinent discussion points will be raised in each experimental write-up. *Brevity in this, and the succeeding, section is a virtue that the writer will want to cultivate.*

7. CONCLUSIONS

This section should not degenerate into a rehash and summary of the entire experimental undertaking, but should succinctly address the results with respect to the Objectives espoused in Section 2. Mr. Franklin would have concluded that lightning was indeed a form of electricity. On the other hand, you might be able to conclude that the stresses at notch tips are proportional to the square root of the notch depth if all other factors are held constant.

In addition to the above formatting, certain guidelines are specific to the work in this class. First, in many industrial situations, management often wishes to check the progress of their researchers. Rather than wading through confusing formats and endless babble, many companies have standardized their laboratory reports on a system of forms, similar to worksheets; whomever is reading the report can easily find what information they need. Thus, in this class, standardized report formats have been prepared.

Other notes:

-Neatness counts; in this regard, nothing has changed since elementary school.
-All figures and tables are to be completely labeled including any relevant units.
-A header appears at the top of each page of the report; it is to be filled out in the following manner:

Tim O'Shenko	January 4, 2009	E Mch 316.5	3/10

-All prelabs are to be completed on an engineering computation pad or on quadrille paper. Chemistry lab notebook paper is unacceptable. Remember that prelabs are due at the start of each lab.

In order to add some "real" flesh to the cryptic outline above, a **Sample** laboratory assignment sheet and experimental report is included for your reading pleasure on the following pages.

Truly Nerdy Diversion: All aficionados of solid mechanics will notice a certain, shall we say, play on words somewhere on this page. A hint to this truly exciting mystery that will no-doubt keep you up at night is, think *"Theory of Elasticity."*

E Mech 316
Lab #0

MATERIALS WITH DIFFERENT TENSILE AND COMPRESSIVE BEHAVIORS

<u>Submitted to:</u>
Esteemed T.A.

Department of Engineering
Science and Mechanics

The Pennsylvania State University

<u>Submitted by:</u>
Stew Dent
January 4, 2009

OBJECTIVE:

Many engineering materials respond quite differently to compressive

than to tensile stresses. Therefore, the stress-strain characteristics

of Plexiglass, an acrylic plastic, will be measured in both tension and

compression, such that a comparison can be made between the two

types of response.

PROCEDURE:

Two tests to failure were undertaken. First, one compression and one tension specimen's diameter and gauge length were each accurately measured. They were both placed in the appropriate grips and tested in an ATS 900 Universal Testing machine; a 0.5 in/min. (1.27 cm/min.) cross head speed was used for both tests. Elongation and load measurements were then recorded.

DATA AND RESULTS:

Table 1. Tensile specimen dimensions.

Gauge Length	0.178 m
Diameter	0.0191 m

Table 2. Compressive specimen dimensions.

Gauge Length	0.102 m
Diameter	0.0508 m

Table 3, Table 4 Data from testing equipment.

Load (N)	Elongation (m)

ANALYSIS AND CALCULATIONS:

•Area Calculation

$$A = \pi r^2$$
$$A_T = \pi (0.0191m)^2$$
$$A_T = 2.87 \times 10^{-4} m^2$$
$$A_c = \pi (0.0508m)^2$$
$$A_c = 2.03 \times 10^{-3} m^2$$

| $A_T = 2.87 \times 10^{-4} m^2$ |
| $A_c = 2.03 \times 10^{-3} m^2$ |

•Sample Stress-Strain Calculation

$$\sigma = \frac{P}{A} \quad \varepsilon = \frac{\Delta L}{L} \quad etc.$$

•Determination of Moduli of Elasticity (E)

$E \equiv$ slope of elastic portion of stress strain curve graphically:

$$E_T = \frac{65MPa}{0.005m/m} = 13GPa$$
$$E_c = \frac{80MPa}{0.00125m/m} = 64GPa$$

| $E_T = 13GPa$ |
| $E_c = 64GPa$ |

•Determination Ultimate Strengths

graphically:

$$\sigma_{UT} = 65 \ MPa$$
$$\sigma_{UC} = 110MPa$$

| $\sigma_{UT} = 65 \ MPa$ |
| $\sigma_{UCS} = 110 \ MPa$ |

•Determination of Strength Ratio, K

$$K = \frac{\sigma_{UC}}{\sigma_{UT}} = \frac{110MPA}{65MPa} = 1.69$$

K=1.69

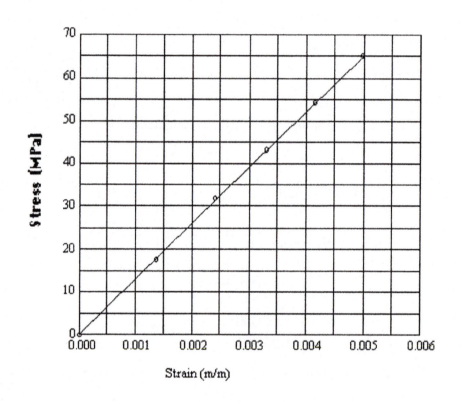

Figure 1. Tensile stress-strain response of plexiglass.

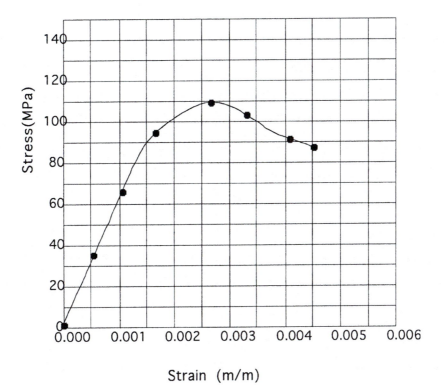

Figure 2. Compressive stress-strain response of plexiglass.

DISCUSSION OF RESULTS:

1. What differences are illustrated between the tensile and compressive mechanical response of plexiglass?

The ultimate strength of the plexiglass in tension is about half of the ultimate strength in compression. Plexiglass acts as two different materials, in compression, and in tension; even the modulus of elasticity was higher in compression than in tension. In tension, plexiglass is a brittle material, while in compression it is a somewhat ductile material.

CONCLUSIONS:
    Plexiglass is indeed a material that has quite different
tensile/compressive responses. Its compressive modulus of elasticity
(measured in compression) is five times greater than in tension. The compressive
ultimate strength is larger than in tension. In tension the material
fails in a brittle manner, while in compression the material fails in a
compressive manner.

DETERMINATIONS OF STRESS
CONCENTRATIONS VIA PHOTOELASTICITY

INTRODUCTION

Determining stresses in a structure experimentally is a venerable approach to that critical part of the design process. At one time, experimental methods were often the only realistic hope to solve for the stress distributions in complex structural elements, as mathematical analyses could not be easily done on such geometries. Nowadays, sophisticated numerical techniques such as finite-element analysis are available to all, and offer the promise of solutions to virtually any problem provided sufficient computational muscle is in hand. Despite the near infatuation that some designers have for computer modeling, the need to physically verify that the stresses are actually what the computers says they are is as great as ever. Indeed, the acceptance of machine-generated design analyses without critical verification continues to result in engineered disasters of various magnitudes. Hence, never just say that "that is what the computer told me." Trust but verify should always be the guiding wisdom for any engineering analysis! Always remember Segall's not so famous equation:

Most Advanced Computer Codes + Junk In = Absolute Junk Out

In particular, structural regions with sharp geometry changes, those features we know as stress concentrators, are the most difficult to numerically analyze with finite-element methods. At the same time, stress concentrators are often the most relevant feature of a design stress analysis, particularly if the material is of low ductility or the service environment features fatigue loading or aggressive chemical species.

Fortunately, *Photoelasticity* allows one to measure the stresses at any point within a member, including regions of high stress concentration. The method is based on the physical phenomenon of certain transparent solids taking on interpretable light patterns when stressed if the light they are transmitting is polarized. As such, the method is somewhat dependent on materials properties of the model. You may fondly recall that your stress analysis formulae are generally material insensitive for elastic stresses, such that the flexural stress in a beam is Mc/I, regardless of the beam material.

The technique is implemented by building a planer model of the structural element, loading the model as in service, and viewing and interpreting the resultant light patterns. In this particular experiment, a circular polariscope (consisting of a light source, two polarized plates, two quarter wave plates, and in some cases, a camera) will be used to demonstrate photoelasticity.

A polarizing sheet or plate transmits only the portion of a light wave whose amplitude vector is parallel to the polarizing direction. A simple polarizing sheet is shown in Figure 1-1. When the polarizer's and the analyzer's (another polarizing sheet) axes are at right angles, all light is extinguished and is appropriately called the 'dark field' configuration. In contrast, when the polarizer and analyzer are aligned such that the axes are parallel, the configuration is now termed "light field."

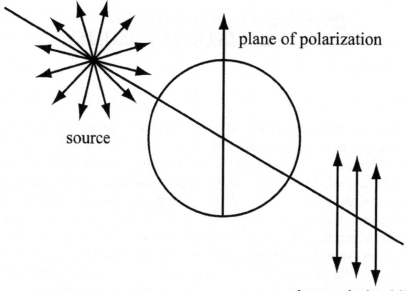

source

plane of polarization

plane polarized light

Figure 1-1. A Circularly shaped plane polarizing sheet.

Wave plates resolve the electric field of the incident light vector into two perpendicular components with different velocities. Therefore, the light emerges from the wave plate with a *positive* phase change. Specifically, a quarter wave plate retards the light such that there is a 90° phase change. Figure 1-2 shows circularly polarized light, as well as the result of incident light passing thorough a polarizing sheet and a quarter wave plate.

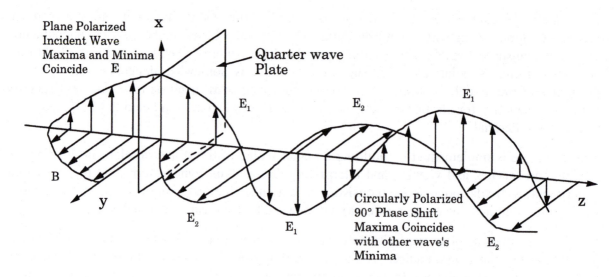

NOTE: Resultant B wave not shown

2

Figure 1-2. Circularly polarized light; note that the phase change is 90°.

A circular polariscope is shown below in Figure 1-3 and consists of all of the before mentioned components including the polarizers, quarter wave plates, and the stressed model.

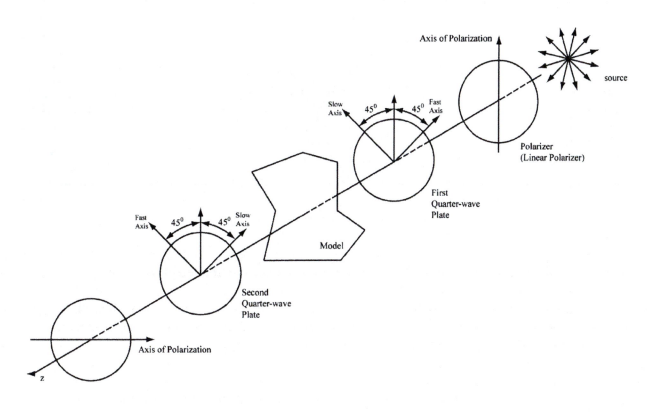

Figure 1-3. A circular polariscope in the "dark field" configuration.

Typical models are usually flat or planer pieces of a transparent material that become temporarily *birefringent* when strained. A birefringent material breaks up polarized light's electric field into two perpendicular vector components; in this case, one along each principal stress (σ_I and σ_{II}) direction. Along one of the principal axes, the speed of the light wave is retarded as it passes through the thickness of the model. Assuming plane stress, the relative retardation is proportional to the difference in the principle stresses and is given as:

$$(\sigma_I - \sigma_{II}) = \frac{CN}{t} \qquad (1\text{-}1)$$

$C \equiv$ Stress-optical coefficient (material property)

$N \equiv$ number or wavelengths of relative retardation (fringe order)

$t \equiv$ model thickness

If the relative retardation is an integral number of wavelengths, extinction occurs and a dark purple fringe can be seen. These fringes are called "tints of passage" while non-integral fringes appear as colors of the spectrum. For example, Figure 1-4 shows tints of passage in a decidedly (and unfortunately) monochromatic fashion.

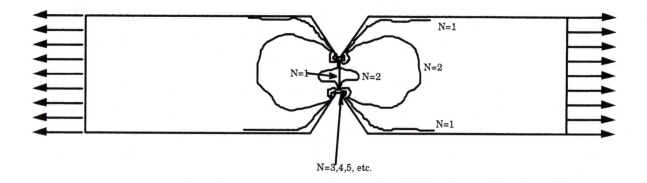

Figure 1-4. A sample notched model exclusively showing integral fringes.

The fringes can be numbered by their growth. For example, as the stress was increased, fringe #1 would move and replace or become fringe #2, and a new fringe would appear to replace the original fringe #1. Table 1-1 converts colors into fringe order (N).

Table 1-1. Approximate color fringe chart; Asterisks mark tints of passage

Fringe Order	Color	Fringe Order	Color
0.0*	black	2.2	blue-green
0.28	gray	2.3	sea green
0.6	yellow	2.4	bright green
0.79	orange	2.5	green-yellow
0.90	red	2.67	pink
1.0*	violet (indigo)	2.7	crimson
1.2	blue-green	2.8	dull purple
1.3	green	2.9	violet-gray
1.38	green-yellow	3.0*	dull gray-green
1.62	orange	3.2	bright gray-green
1.81	red	3.5	light green
2.0*	violet (indigo)	3.6	pink

4

As would be expected, some photoelastic materials are more sensitive than others and produce more fringes under a given stress. For instance, ordinary window glass has such a low sensitivity that it is impractical to use for modeling. This is truly unfortunate since a higher sensitivity would allow one to easily gauge the stresses in buildings, aircraft, or any windowed structure. Table 1-2 lists some typical values. Note that lower values of the stress-optical coefficient correspond to higher sensitivity.

Table 1-2. Typical stress optical coefficients.

Material	Stress-optical coefficient, C ($\frac{N}{m \cdot fr}$)
Gelatin	30
Polyurethane Rubber	170
Epoxy Resins	10500

At the load free boundaries of a model, the principal directions are known, and the normal stress perpendicular to the boundary is zero, so equation (1-1) becomes:

$$\sigma_I = \frac{CN}{t} \qquad (1\text{-}2)$$

Within the model, the lines of maximum shear stress in the plane of the model are coincident with the observed fringes because:

$$\tau_{max} = \frac{\sigma_I - \sigma_{II}}{2} \qquad (1\text{-}3)$$

Therefore

$$\tau_{max} = \frac{CN}{2t} \qquad (1\text{-}4)$$

These fringes, or lines of constant maximum shear stress, are called isochromatics. If the quarter wave plates are removed, black lines called isoclinics appear wherever the direction of the principal stresses were coincident with the axis of the polarizer. Any polariscope without quarter-wave plates is called a plane polariscope.

The experiment you are no-doubt eager to start will use a standard circular polariscope in the dark field arrangement to observe isochromatic fringes. Two types of specimens (all made of PSM-5 epoxy) will be used for the experiment; a calibration specimen will be used to determine the stress-optical coefficient while notched configurations with various notch depths and radii of curvature will be used to investigate stress concentrations. The load will be applied using dead weights on a frame that features a simple lever for magnification.

PROCEDURE

- Measure and record all relevant dimensions of the loading frame

- Measure and record the width and thickness of each specimen. Use the metric micrometer whenever possible. The notch depths (a) and radii of curvature (ρ) are already recorded, as they would be difficult to measure with calipers.

- Set up the polariscope and verify that it is in the dark field configuration. *Not So Subtle Hint:* the relevant axes are marked on the polarizers and wave plates.

- Place a specimen in the loading frame with the loading pins.

- GRADUALLY apply the dead weight load. DO NOT allow the weights to hang on the specimens for more than a minute or two as extended loading will cause permanent (and expensive) deformation (see Lab # 5).

- Relieve the load by lifting up on the lever until there is no load. Slowly reload the specimen. During reloading, observe the changing fringes. The fringes should ripple through the color fringe chart (Table 1-1).

- Visually estimate and record the highest fringe order (N_{max}) to the nearest tenth.

- Sketch each specimen tested by your group. Label the fringes with numbers and colors.

- Repeat as necessary.

ANALYSIS OF DATA

- Determine the unnotched area of the specimens and show one example calculation.

- Determine the actual applied load.

- Determine the far field stresses (the stresses away from the notches) by dividing the load by the unnotched area.

- Complete table 2.

- Calculate the stress optical coefficient from the calibration specimen data. If your group does not have the calibration specimen, obtain the necessary values from the appropriate group.

- Calculate the maximum stresses using equation (1-2).

- For each notched model, determine the stress concentration factor. Recall:

$$K_\sigma = \frac{\sigma_{max}}{\sigma_0}$$

- Put your values on the board and complete table 3.

- Plot the values of K vs. $\sqrt{a}$ and K vs. $\sqrt{\frac{1}{\rho}}$. There should be four data points per plot.

- Determine the constants of proportionality (C_d and C_n) from the slope of the plots. Note the definitions of C_d and C_n:

$$K_\sigma = \frac{\sqrt{a}}{C_d} \qquad\qquad\qquad K_\sigma = \frac{C_n}{\sqrt{\rho}}$$

1.   Define the following terms:

1. Birefringent

2. Photoelastic

3. Isochromatic

4. Isoclinic

2.   A photoelastic experiment was undertaken using an unknown type of rubber. The following data was collected:

Calibration specimen
Width = 40 mm
Thickness = 5mm
Load = 27.54N
max fringe color observed: orange red

Find the stress optical coefficient for this material.

3.   A notched material was also made from this unknown material. Assuming that the cross-sectional area away from the notch is the same as that of the calibration, find the stress concentration factor.

Load = 19.5N
max fringe: violet gray (2.9)

E Mech 316
Lab #1

**DETERMINATION OF STRESS
CONCENTRATIONS VIA PHOTOELASTICITY**

Submitted to:

Department of Engineering
Science and Mechanics

The Pennsylvania State University

Submitted by:

	10		/

OBJECTIVE:

_____

_____

_____

_____

_____

_____

_____

_____

_____

_____

_____

			/

PROCEDURE:

_____

_____

_____

_____

_____

_____

_____

_____

_____

_____

_____

_____

DATA AND RESULTS:

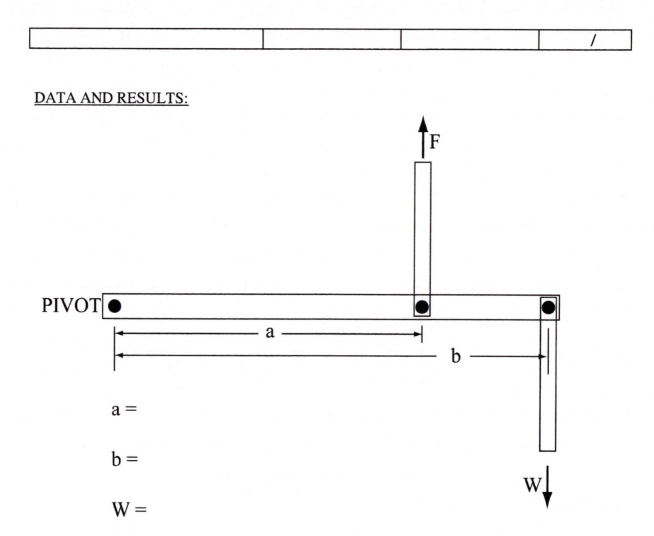

a =

b =

W =

Figure 1. Loading frame with relevant dimensions.

Table 1. Dimension and fringe data.

Specimen	Width (mm)	Thickness (mm)	N (fr)

			/

Figure 2. Sketch of Specimen

Figure 3. Sketch of Specimen

			/

ANALYSIS AND CALCULATIONS:

•Determination of Areas (sample calculation only):

•Determination of actual applied load, F:

•Determination of far field stresses, σ_0:

•Determination of Stress-Optical Constant, C:

			/

•Determination of maximum stresses, σ_{max}:

•Determination of Stress Concentration Factors, K_{σ}:

•Determination of Constants of Proportionality, C_d & C_n:

			/

Table 2. Our groups values.

Specimen	Load (N)	Area (m^2)	Stress, σ_0 (MPa)

Table 3. Stress concentration values.

Specimen	Load (N)	a (mm)	ρ (mm)	$\sqrt{a}$ ($\sqrt{m}$)	$1/\sqrt{\rho}$ ($1/\sqrt{m}$)	σ_0 (MPa)	N_{max} (fr)	σ_{max} (MPa)	K_σ
C									
1		13	0.79						
2		9.5	0.39						
3		9.5	0.79						
4		9.5	1.6						
5		9.5	3.2						
6		6.4	0.79						
7		3.2	0.79						

Figure 4.

Figure 5.

			/

DISCUSSION OF RESULTS:

1. How does your calculated value of the stress-optical coefficient of PSM-5 epoxy compare with other photoelastic materials? Is this value expected?

_____

_____

_____

_____

_____

_____

_____

2. Name and define the primary material property that allows certain materials to be used in photoelastic stress analysis.

_____

_____

_____

_____

_____

_____

3. Under a constant load, how would the number of fringes vary if the thickness of a PSM-5 epoxy model is increased? Explain your response analytically.

		20		/

4. Assuming the load were increased, where would the notched specimens fail first? Why?

_____

_____

_____

_____

_____

_____

_____

_____

			/

CONCLUSIONS:

_____

_____

_____

_____

_____

_____

_____

_____

_____

_____

_____

_____

LAB #2.

STRESS ANALYSIS VIA
STRAIN GAUGE ROSETTES

INTRODUCTION

As you will no-doubt discover once you enter the venerable "Real World," not all load configurations are as simple as textbook examples such as pure tension or pure bending. In fact, most loading schemes have complex multi-axial stresses. While many engineers have begun to rely on computer codes for such complicated analyses, empirical solutions are still very important, especially for verification as discussed earlier.

Most experimental stress-strain data are obtained from some form of elongation and/or strain measurements since these are entities easily measured. Optical techniques to measure deformation include polariscopes (see Lab #1), interferometers, optical strain gauges, as well as a variety of methods using mirrors to magnify the deformation through projected images; extensometers are the most common mechanical method for such measurements. However, electrical resistance strain gauges are the devices most frequently used (preferred by 9 out of 10 professional Stress-Guessers) in experimental stress analysis of structures, especially when approximating elongation at a point.

Electrical strain gauges come in a variety of forms including capacitance gauges inductance gauges, and resistance gauges. Clearly, the resistance strain-gauge is the preferred device as it is based on a simple and easily implemented principle, namely the resistance of a metal conductor changes when strained. Interesting, this property was originally observed by the prolific Lord Kelvin in 1856 when he measured the changing resistance of iron and copper wires hanging in tension. In his experiment, he correctly concluded that the changes in resistance of the wire was a function of strain, e.g:

$$\varepsilon = \frac{\Delta L}{L} = k\frac{\Delta R}{R} \qquad (2\text{-}1)$$

In addition, he also correctly surmised that the change in resistance is material sensitive and that a Wheatstone bridge can be used to accurately measure the resistance changes provided accurate resistors are in hand. Furthermore, it was found:

$$\Delta\varepsilon = f\frac{\Delta R}{R} \qquad (2\text{-}2)$$

where f is the "gauge factor." The gauge factor can vary from about 1 to 3; most commercial gauges have a value of equal to around 2.0.

23

Commercial gauges were initially made of wire wound back and forth and fixed on a piece of tissue paper or encapsulated in a polymeric material. The winding is necessary to obtain sufficiently high resistance and maintaining practically sized gauge lengths, while the covering is for protection. Today, most gauges are stamped from thin foils by precision dies onto a very thin plastic base.

Strain gauges can only be attached to the surface of material; thus, the gauges can only measure strains in the plane of that surface. Given this limitation, gauges are usually affixed to the surface with commercial adhesives such as epoxy, Superglue, and even common model glue. If needed, specialized epoxies are readily available for intricate or delicate gauges.

Fortunately, surface measurements are clearly desirable as critical stresses and strains generally occur there. For example, a beam subjected to bending enjoys a maximum stress at the furthest distance from the neutral axis (surface), usually referred to as the outer fiber. At the surface, a state of plane stress exists ($\sigma_3=0$). Thus, the state of stress can be completely defined by the quantities, σ_{xx}, σ_{yy}, and τ_{xy}. Three independent measurements, ε_{xx}, ε_{yy}, and γ_{xy}, must therefore be made to calculate these three quantities (3 equations and three unknowns); these quantities are derived from the data of three non-collinear strain gauges.

Groups of strain gauges on a single backing are called *strain rosettes* or *strain gauge rosettes*. Rosettes are commonly used because the angular relationship between the gauges is precisely known; the two most common are the rectangular and the delta rosettes pictured below in Figure 2-1.

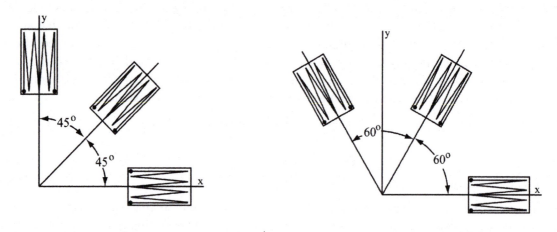

Rectangular Rosette   Delta Rosette

Figure 2-1. Rosette schematics showing two common configurations.

Rosette usage requires that one principal direction be perpendicular to the plane of the rosette. Fortunately, this is always true on a free surface, while the other two principal directions lie in the plane of the rosette, but not necessarily in the direction of the gauges. Thus, the principal strains or stresses can be calculated using geometry and everyone's favorite, Mohr's circle.

24

The present experiment will address the stress analysis of a thin-walled pressure vessel. Pressure vessels are commonly found in all sorts of engineering situations such as nuclear and fossil fuel power generation, petrochemical processing, as well as water and natural gas distribution systems to name just a few. Pressure vessels can have complex stress distributions, especially if they are thick-walled, reinforced with rings, supported by braces, subjected to varying temperatures, and/or buried underground. However, if plane-stress conditions in the wall of the vessel (away from the ends) is assumed, the classic "Strength of Materials" stress solution for a closed-end, internally pressurized thin-walled cylindrical pressure vessel for the hoop, tangential or circumferential stress, σ_θ is given as:

$$\sigma_\theta = \sigma_I = \frac{p\,\bar{r}}{t} \tag{2-3}$$

for the σaxial or longitudinal stress, the solution is

$$\sigma_{ax} = \sigma_{II} = \frac{p\,\bar{r}}{2t} \tag{2-4}$$

where $\bar{r}$ is the vessel mean radius, t is the vessel wall thickness, and p is the gauge pressure. These solutions have been found to be valid for pressure vessels where the thickness is small as compared to the radius as defined by:

$$\frac{\bar{r}}{t} \geq 10 \tag{2-5}$$

This lab will use rectangular strain-gauge rosettes on an aluminum pressure vessel to compare three approaches to analyzing the closed-end, cylindrical pressure vessel, namely the analytical strength of materials solution, the experimental solution obtained from a gauge aligned with the principal axes, and the solution from a randomly oriented rosette. Pictured below are the approximate orientations of the rosettes on the thin-walled pressure vessel.

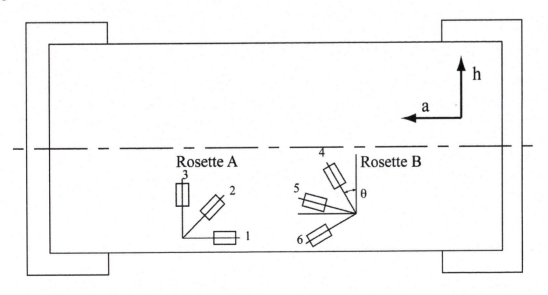

Figure 2-2. Approximate orientations of rosettes on the thin-walled pressure vessel.

PROCEDURE

- Referring to Figure 2-3 below, open all of the valves on the console.

- Unplug the strain indicator box and turn it to "µε." Turn the dial readout to zero. Verify that there is no internal pressure by double checking the pressure gauge. Zero each channel (1-6) on the balance with a screw driver. Record your vessel number in the heading of Table 2; some other relevant data appears in Table 1.

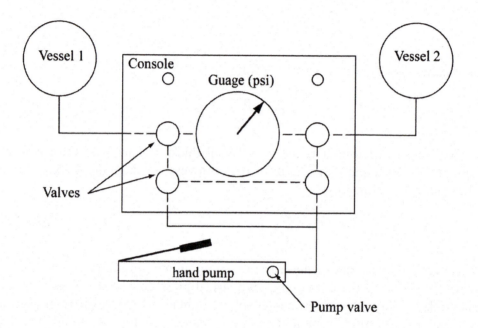

Figure 2-3. Vessels and strain-guage console.

- Wait for the other group(s) to balance their gauges.

- Close the pump valve and pump to approximately 1.5 MPa (220 psi). *DO NOT ALLOW THE PRESSURE TO EXCEED 1.7 MPa (250 psi) OR ELSE YOUR DAY WILL BE RUINED!* Have one group member watch the pressure gauge for possible losses. Record this pressure.

- Take the six strain readings. Do not disturb the balance knobs during the experiment. To take a reading, turn the dial until the meter balances. Record the strain, switch channels, and repeat as necessary.

- After all groups are finished, evacuate the vessels with the valve on the pump.

- Turn off the indicator and plug it back into the outlet.

LAB #3.

TENSILE BEHAVIOR OF A DUCTILE METAL

INTRODUCTION

One of the oldest methods of quantitatively assessing the mechanical behavior of materials is the tension test, a state of loading easily and incontestably achieved with a minimum of technical sophistication. Natural philosophers of the Renaissance performed such tests by hanging filaments from cathedral ceilings (presumably, in the off hours) and proceeded to probe the deformation characteristics of various metals and the main filamentary polymer of the day, catgut.

Smelly tests and poor kitty notwithstanding, modern engineers still use the tension test frequently, but with less odorous materials. For example, it is used as a quality control measurement involving large numbers of specimens; tension tests are also used in industry where new and novel materials (synthesize catgut perhaps?) are created.

The most common representation of the load/deformation characteristics of a ductile material is obtained by the use of the *nominal* stress and strain parameters to represent load and deformation, respectively:

$$\sigma = \frac{P}{A_o} \qquad \text{Nominal Tensile Stress} \qquad (3\text{-}1)$$

$$\varepsilon = \frac{\Delta L}{L_o} \qquad \text{Nominal Tensile Strain} \qquad (3\text{-}2)$$

wherein P is the load in pounds or Newtons, A_O is the unloaded cross sectional area of the gauge length, L_O which elongates by ΔL under the influence of the load. A typical nominal tensile stress-strain curve for a ductile material is shown on the next page in Figure 3-1.

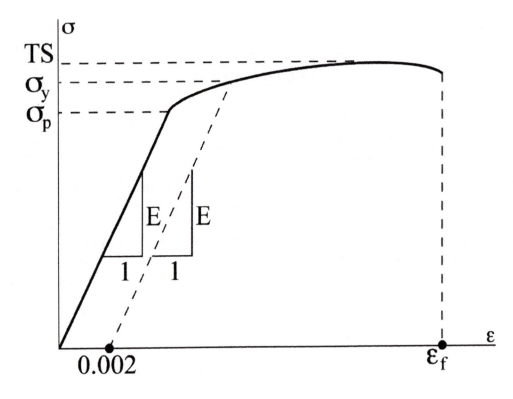

Figure 3-1. "Typical" stress-strain curve.

Elastic parameters found are the proportional limit, σ_p, and (Young's) modulus of elasticity, E. Additionally, strength parameters are the 0.2% yield strength, σ_y, usually obtained by graphical construction, and the (ultimate) tensile strength, TS or σ_u. The ductility measure is the % Elongation or 100 times the strain at failure, 100 x ε_f. Derived work (energy) absorption measures are the modulus of resilience, the triangular area under the curve up to the proportional limit and the toughness defined as the area under the entire curve to fracture; the latter can be approximated by the modulus of toughness, the average of the yield and tensile strengths, times the strain to fracture.

In this lab, the mechanical responses of a highly ductile metallic alloy will be quantitatively determined and compared to published data. As such, Magnesium and Silicon alloyed Aluminum specimens, commonly referred to as Al 6061-T6, will be tested in uniaxial tension to failure. For the tests, the classically shaped cylindrical dog-bones specimens will be used. Either a hand cranked or a simple automatic testing machine will be used to generate load-elongation data.

During the tests, deformations will be measured via a mechanical extensometer, which consists of a knife edge affixed to the specimen and holding a dial gauge which measures the relative separation distance to another independent knife edge as shown in Figure 3-2.

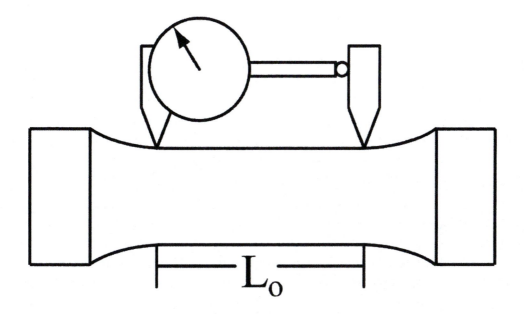

Figure 3-2. Dog-bone specimen with an extensometer and gauge length indicated.

The *Metals Handbook* (Vol 2., Ninth ed.) reports the following mechanical responses:

Table 3-1. Typical properties of Al 6061-T6

Property	Magnitude (MPa)
Young's Modulus, E	68300
Yield Strength, σ_y	276
Ultimate Strength, σ_u	310

PROCEDURE

- Measure the diameter of the dog bone specimen at the middle of the reduced section using the metric micrometer.

- Load the specimen into the testing machine grips so that at least 3/4 of the gripping surface is in contact with the specimen.

- Attach the extensometer in its fully compressed position to the sample. ONLY TIGHTEN THE THUMBSCREWS FINGER TIGHT. Tightening the screws too much will pit the specimen; the resulting indentation will act as a stress concentration and the specimen could fail prematurely.

- Using the dividers, measure the distance between the knife edges on the extensometer; this is important as it is the gauge length.

- Load[*] the specimen to approximately 200lb (1000N) to make sure the extensometer is functioning.

- Unload, then begin the test, taking the first load up to about 10% of the expected yield; definitely plan on recording loads and deformations in about 8 equal increments to the expected yield. Thereafter, elongations will increase much faster than loads, so you should plan on 8 to 10 readings before expected fracture.[†] Record this data as indicated. NOTE: As the test progresses, periodically re-tighten the thumb screws on the extensometer to compensate for the specimen thinning as it elongates (the Poisson ratio effect). Be sure to record the relevant quantities in table 2.

- Remove the fractured specimen from the grips and measure the final diameter.

- Draw a sketch of the failed specimen. Include details regarding the fracture surface.

---

[*] More specific instructions regarding the operation of the equipment will be given in class.
[†] See pre-lab calculations for a guide.

ANALYSIS OF DATA

- Calculate the specimen's area. Show a sample calculation for stress and strain and complete Table 3.

- Plot the entire stress-strain curve. Use as much of the scale as possible to maximize data resolution. You may wish to use the long axis (long side of the paper) for the strain axis.

- Plot another stress-strain curve showing the linearly elastic portion of the previous plot. The strain axis must be long enough so that the 0.2% offset yield strength can be drawn and determined accurately.

- Determine the indicated material response parameters

1. Read over the 'ANALYSIS AND CALCULATIONS' section of the lab report. Familiarize yourself with these quantities. You may wish to bring along a text for reference.

2. Table 3-1 is repeated here for your pleasure and convenience:

Table 3-1. Typical properties of Al 6061-T6.

Property	Magnitude (MPa)
Young's Modulus, E	68300
Yield Strength, σ_y	276
Ultimate Strength, σ_u	310

-   Assume a diameter of 0.25 in.(635 cm) and a gauge length of 2.5 in.(6.35 cm).

-   Calculate 10% of the yield strength. What is the required load? (Answer in both S.I. and English units) Round this value to the nearest 50 pounds or 100N.

-   If this particular specimen were completely elastic to yield, calculate about 8 evenly spaced loads between the 10% of yield load and the yield load. (Again, answer in both S.I. and English units and round these values to the nearest 50 pounds or 100N.

-   Calculate the difference between the strain at failure (let $\varepsilon_f=0.08$) and strain at yielding. Convert this to a change in elongation. Divide by 10 and set this equal to ΔL_{inc}. Calculate ΔL_{inc} in English units only.

$$10\% \, \sigma_y = 27.6 \qquad \sigma = P/A_o$$
$$A_o = \frac{\pi D^2}{4} = \frac{\pi (.635)^2_m}{4} = .3167 \, m^2$$

LAB #4.

DETERMINATIONS OF
NOTCH SENSITIVITY
IN LOW DUCTILITY MATERIALS

INTRODUCTION

Sudden changes or discontinuities in the geometry of a structural member usually give rise to locally increased values of stress and strain at that point. The most common example is a small hole in a large (relative to the size of the hole) tensile specimen as shown in Figure 1.

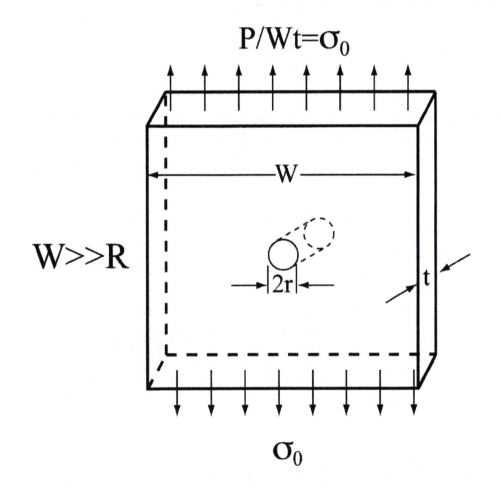

$$P/Wt = \sigma_0$$

$$W \gg R$$

$$\sigma_0$$

Figure 1. Relatively wide rectangular plate with cylindrical hole.

At first glance, one might assume that the tensile stresses near the hole would simply be determined by the load P that causes the far field stress, σ_0, divided by the effective area defined as $A_{eff} = t \cdot (W - 2r)$. While simple pleasures are always the best, no such luck in this situation. In fact, the magnitude of the stresses near the hole are much greater than the intuitive P/A_{eff} stress approach would suggest. On the other hand, the stresses in regions sufficiently far from the hole are unaffected by it, a clear demonstration of the *Principle of St. Venant.*

The principle of St. Venant was promulgated in 1855. Simply stated, at a characteristic distance away from the discontinuity, i.e. the hole in this case, the stresses are unaffected by the geometrical changes or discontinuities. In other words, at this distance from the hole, the difference between the actual stress and the far field stress is negligible. Since analytical solutions are not always feasible for discontinuities in finite bodies, a wide collection of experimentally (many using photoelasticity ala Lab # 1) determined results exist for most common situations encountered in engineering practice.

Because maximum stresses are the ones that interest us the most as engineers, the *elastic stress concentration factor*, K_σ, has been defined as:

$$K_\sigma = \frac{\sigma_{MAX}}{\sigma_o}$$

(4-1)

where σ_{max} is the maximum concentrated stress and σ_O is the far field stress that is unaffected by the discontinuity. For the example just given, $K_\sigma = 3.0$, and the stress distribution is shown in Figure 2.

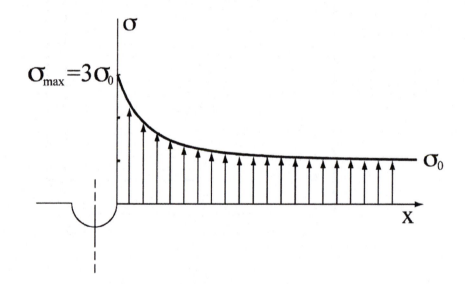

Figure 4-2. Stress distribution and concentration around cylindrical hole.

Simplicity is rarely simple to achieve, so most stress concentrations are not as easy as the above example. In fact, stress concentrations usually occur not only as holes, but also as joints, corners, fillets, and keyways. Moreover, not all discontinuities are the result of conscious design decisions, but may instead be intrinsic to the material; such intrinsic stress concentrators include microscopic voids, inclusions, grain differences to name just a few.

			/

Stress concentrations can have differing effects on many materials and/or in various service contexts. Composite materials will behave very differently when compared to most monolithic materials including alloys and ceramics. In monolithic materials with high ductility, stress concentrators are not usually detrimental to static strength. On the other hand, in service conditions where stresses vary cyclically, the fatigue strength may be strongly influenced by notches and notch-like features.

Low ductility materials are almost always affected by abrupt changes in geometry. In fact, stress concentrators can cause structures to fail well below their predicted ultimate strength when these strengths are measured in notch-free samples. The decrease in the "apparent" ultimate strength is most acute in brittle materials and would be expected to be a function of the stress concentration factor.

The present experiment will compare notch sensitivity in 7075-T6 wrought aluminum, a low ductility, high strength alloy, with a medium strength, medium ductility aluminum (6061-T6). The former alloy is used in ski poles, aircraft structures and other highly stressed structural applications where very high strength is needed, the latter in applications where forming ductility may be more important than maximized strength. The *Metals Handbook* (Volume 2, Ninth Edition, ASM, Metals Park, Ohio) reports 572 MPa as 7075-T6's ultimate tensile strength with 11.0% elongation, whereas 6061-T6 reports values of 310 Mpa and 17.0%, respectively. Several specimens with different notch depths will be tested to failure with the experimental strength values compared to those theoretically predicted from the effective ligament area. From this comparison, conclusions can be drawn as to the material's notch sensitivity.

PROCEDURE

- Measure the width and thickness of each specimen; the notch depths will be specified by the lab instructor and used to compute the ligament width. Refer to Figure 3 for definitions of these measures. Record this information in Table 1 and Table 2.

- Place a specimen in the grips of the testing machine, making sure that the specimen is properly aligned with specimen edge parallel to the edge of the jaw. At least 3/4 of the grips should be in contact with the specimen.

- Load the specimen to failure. During loading, observe where the crack initiates. Record the ultimate load in Table 3.

- Remove the fractured specimen and repeat as necessary for the other specimens.

ANALYSIS OF DATA

• Determine the full, or total cross-sectional area of each specimen. Show one sample calculation.

• Derive an equation for the ligament's cross-sectional area, A_{lig}, of each type of specimen. Calculate each A_{lig}, showing one sample calculation of each type. See Figure 3.

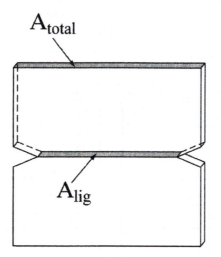

Figure 3. Ligament and total areas.

• Calculate the percent of full specimen area, A%, for each specimen.

$$A\% = \left(\frac{A_{lig}}{A_{total}}\right) \cdot 100 \qquad (4\text{-}2)$$

• Calculate the ligament effective ultimate strength for each specimen:

$$\sigma_{lig} = \frac{P_{fail}}{A_{lig}} \qquad (4\text{-}3)$$

• Complete Table 4 and Table 5.

• Using the definitions just given, plot ligament effective ultimate strength versus percent of full specimen area. Note that the theoretical line will be horizontal and equal to the material's reported ultimate strength if there is no notch sensitivity for the material.

• Modern fracture mechanics suggests a different approach to predicting the strength of notched plates in high strength alloys. Plot, on the same graph as the above correlation, the parameter $P_{fail}(\textbf{notch depth})^{1/2}$ versus percent of full specimen area for the 7075-T6 material only.

63

E Mech 316
Pre-lab #4

1. Hydrogen gas is used in enormous quantities in the Haber process to form ammonia gas:

$$N_2(g) + 3H_2(g) \leftrightarrow 2NH_3(g)$$

The usual problem in Hydrogen processing is from environmental degradation such as hydrogen embrittlement of the pipeline and of the joints. This particular system has specially coated parts to reduce this damage. Unfortunately, catastrophic failures have occurred at the welds, caused by the excessive local stresses due to bending of the pipeline. The project engineer has decided to reduce the local stresses at the weld fillets by increasing their radius of curvature from the present value shown below. If the industrious engineer has decided that the local stresses are to be halved, what should she make the new radii? Hint: recall the proportionality $K \propto 1/\sqrt{\rho}$

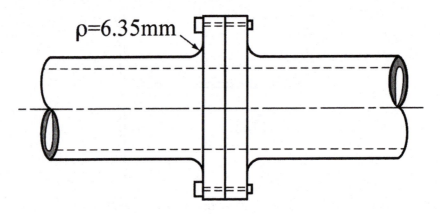

$\rho = 6.35\text{mm}$

2. A crack has been visually observed on a large piece of equipment by a technician on the shop floor. Because of the large shut-down cost (often millions of dollars/day), the engineer must devise a way to temporarily stop, or at least slow, the crack's propagation until a replacement can be made. The crack is in a large steel plate and is shown in an exaggerated form below. Devise a manner to slow the cracks propagation and, thus, minimize down time.

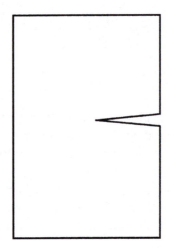

E Mech 316
Lab #4

DETERMINATIONS OF
NOTCH SENSITIVITY
IN LOW DUCTILITY MATERIALS

Submitted to:

Department of Engineering
Science and Mechanics

The Pennsylvania State University

Submitted by:

		66		/

OBJECTIVE:

_____

_____

_____

_____

_____

_____

_____

_____

_____

_____

_____

	67		/

PROCEDURE:

_____

_____

_____

_____

_____

_____

_____

_____

_____

_____

_____

				/

DATA AND RESULTS:

Table 1. Notched 6061-T6 specimen data.

Specimen	width, w ( )	thickness, t ( )	depth, a ( )	ligament width, w_{lig} ( )

Table 2. Notched 7075-T6 specimen data.

Specimen	width, w ( )	thickness, t ( )	depth, a ( )	ligament width, w_{lig} ( )

Table 3. Specimen rupture data.

Specimen	Failure Load, P (N)

			/

ANALYSIS OF DATA

•Determination of specimen full area:

•Derivation and determination of ligament area:

•Calculation of percent of full specimen area:

•Calculation of ligament effective ultimate strength:

				/

Table 4. 6061-T6 Notched specimen analysis results.

Specimen	Total Area, A ( )	A$_{lig}$ ( )	A%	$\sigma_{eff-lig}$ ( )
-	-	-	100.	310

Table 5. 7075-T6 Notched specimen analysis results.

Specimen	Total Area, A ( )	A$_{lig}$ ( )	A%	$\sigma_{eff-lig}$ ( )
-	-	-	100.	572

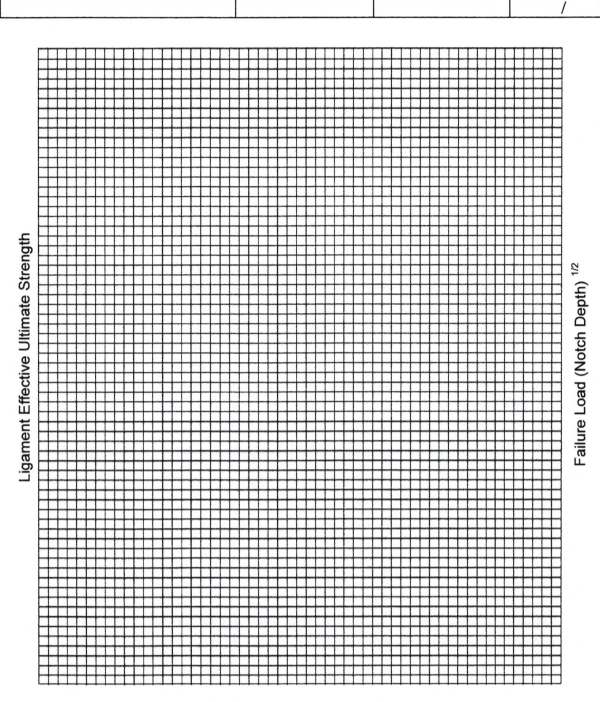

Figure 1

DISCUSSION OF RESULTS

1.   If an alloy were flaw or notch *insensitive*, the ligament effective ultimate strengths would not depend on the effective cross-sectional areas. Is either alloy sensitive to the presence of geometrical discontinuities? Support your response with references to the two sets of data collected.

_____

_____

_____

_____

_____

_____

_____

2.   Based on the results of the fracture tests conducted on the 7075-T6 alloy, does the parameter $P_{fail}(notch\ size)^{1/2}$ appear to correlate both load and notch size simultaneously to describe failure?

_____

_____

_____

_____

_____

_____

_____

3.   Aluminum alloy 1100 is the type used in electrical conductors, very soft but ductile. Based on your experiments, what would your expectations be as to its notch sensitivity

_____

_____

_____

_____

_____

_____

_____

			/

4. Comment on the fracture surfaces. Were they expected for this kind of test?

_____

_____

_____

_____

_____

_____

_____

_____

	74		/

CONCLUSIONS:

_____

_____

_____

_____

_____

_____

_____

_____

_____

_____

_____

_____

LAB #5.

VISCOELASTIC RESPONSE

INTRODUCTION

In realty, the behaviors of most materials are always temperature dependent in some respect. For example, at different temperatures, polymeric materials can be brittle, leathery, rubbery, or even liquid-like (behold, the Terminator); most polymers become brittle like ceramics at cryogenic temperatures. Materials operating at very high temperatures (relative to their absolute melting or transition temperatures: $T > \frac{1}{2} T_{Melt}$ or $T > \frac{1}{2} T_{Trans}$), deform at a *rate* dependent on the stress applied. It is easiest to observe this temperature dependent response in organic materials such as polymers because room temperature is actually "high" for these materials. However, and as stated earlier, all materials exhibit some form of temperature sensitive behavior.

Because the various mechanisms that govern this sort of temperature dependence are extremely complicated, it can be very difficult to understand and predict the mechanical response of materials. Nonetheless, simplified mechanical models can often be devised to describe various behaviors of engineering utility including time dependent behaviors such as viscoelasticity. When viscoelasticity is involved, the simplified methods just mentioned often involve two main elements: a spring and a dashpot. Various combinations of these elements provide a convenient device to derive mathematical expressions and differential equations relating stress, strain and time; these expressions are called the *constitutive equations*.

Any materials exhibiting some form of linear elastic behavior can be described in part, by the simplest of the two elements, the spring. In this case, the force on a spring is a function of the displacement and the spring constant as shown in Figure 1.

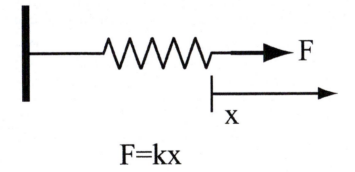

Figure 5-1. Spring element.

For a force applied to a unit area, the spring equation becomes Hooke's Law for tension:

$$\sigma = E\varepsilon \qquad\qquad (5\text{-}1)$$

On the other hand, the time dependent behavior or viscous flow of a material can be described by the second element, the simple dashpot. A dashpot is nothing more than a piston in a fluid filled cylinder; examples include auto shock absorbers and storm door cylinders that prevent them from slamming (although these devices also include springs. Unlike the spring, the dashpot requires time to deform as governed by the time for the liquid to flow past the piston. A dashpot schematic is shown below in Figure 5-2.

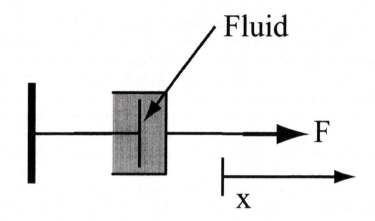

Figure 5-2. A simple dashpot element.

If the cylinder is filled with an ideal Newtonian fluid, the deformation rate is proportional to the force transmitted:

$$F = \beta \frac{dx}{dt} \qquad (5\text{-}2)$$

where β is a constant related to the viscosity of the Newtonian fluid. When the force is applied over a unit area and the deformation over a unit length, Equation 5-2 becomes:

$$\sigma = \eta \frac{d\varepsilon}{dt} \qquad (5\text{-}3)$$

where η represents the viscosity that is a measure of material resistance to deformation with time; the units for viscosity are stress-time. Viscosity is clearly temperature dependent and is governed by an Arrhenius-type equation.

Despite the simplicity of the analysis just used, the material behaviors of many materials can be modeled by a combination of these two elements. One of the simplest is the Voight or Kelvin model shown below under an applied stress.

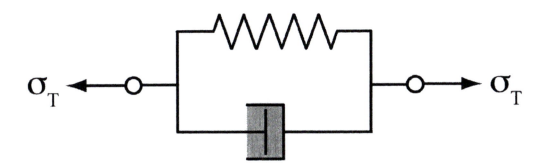

Figure 5-3. Voight or Kelvin Model with applied stress, σ_T.

Because the elements are in parallel, any deformation or strain in the total system is equal to the deformation or strain in each of the single elements.

$$\varepsilon_T = \varepsilon_S = \varepsilon_D \qquad (5\text{-}4)$$

where the subscripts T, S, and D stand for total system, spring, and dashpot, respectively. A free body diagram of the Voight model is shown below n Figure 5-4.

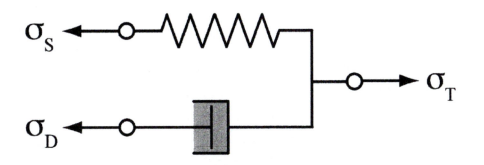

Figure 5-4. Free body diagram of the Voight model.

Usually, stresses *cannot be added algebraically*, but in this case the areas are all unity and the stresses are all in the same direction; therefore, equation 5-5 is obtained.

$$\sigma_T = \sigma_S + \sigma_D \qquad (5\text{-}5)$$

Substituting the spring and dashpot constitutive equations, 5-1 and 5-3, and noting the strain equivalence described by Equation 5-4, yields the following differential constitutive equation:

$$\sigma(t) = E\varepsilon + \eta\frac{d\varepsilon}{dt} \qquad (5\text{-}6)$$

Examining equation 5-4, one can see that the Voight model cannot instantly deform (an elastic response common to all materials), since the dashpot takes time to move. As such, the model is somewhat limited in its practical application.

Another simple model used in the analysis of viscoelasticity is the Maxwell model, shown below with the same applied stress:

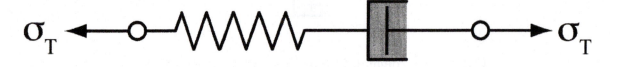

Figure 5-5. Maxwell model with an applied stress, σ_T.

Since the elements are connected in a series configuration, equilibrium demands:

$$\sigma_T = \sigma_S = \sigma_D \qquad (5\text{-}7)$$

The total displacement or strain will be the sum of the individual displacements or strains, such that:

$$\varepsilon_T = \varepsilon_S + \varepsilon_D \qquad (5\text{-}8)$$

Differentiating equation 5-8 with respect to time yields the following first-order differential equation in terms of strain:

$$\frac{d\varepsilon_T}{dt} = \frac{d\varepsilon_S}{dt} + \frac{d\varepsilon_D}{dt} \qquad (5\text{-}9)$$

Substituting the spring and dashpot constitutive equations, 5-1 and 5-3, and noting the stress equivalence defined by Equation 5-7 yields the first order separable differential equation:

$$\frac{d\varepsilon}{dt} = \frac{1}{E}\frac{d\sigma}{dt} + \frac{\sigma}{\eta} \qquad (5\text{-}10)$$

One structural application to which the Maxwell model can indeed be applied is stress relaxation; when a material is subjected to a constant strain, the stress in the material is empirically observed to decrease over time through a process known as stress relaxation. An example of stress relaxation is a pre-stressed rivet or bolt loosening with time. For anyone who ever played with a plastic erector set, you experienced this as the plastic bolts would always loosen over time regardless of how tightly the bolt was turned. In most stress relaxation experiments, the specimen is assumed to instantaneously deform, this deformation is then considered fixed, and the stress is monitored over time. The resulting stress and strain history is shown in Figure 5-6.

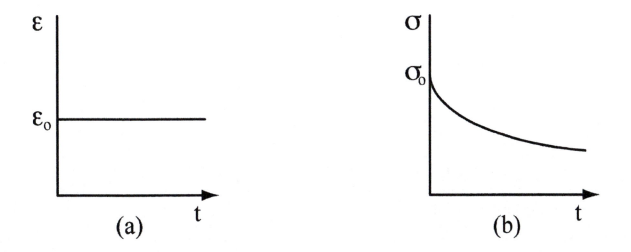

Figure 5-6. (a) Imposed strain history and (b) resultant stress history.

Since the material is under a constant strain, Equation 5-10 becomes:

$$0 = \frac{1}{E}\frac{d\sigma}{dt} + \frac{\sigma}{\eta} \qquad (5\text{-}11)$$

Separating variables, integrating, and finally simplifying to:

$$\frac{d\sigma}{\sigma} = -\frac{E}{\eta}dt \qquad (5\text{-}12)$$

results in the following relationship showing an expected exponential time dependence

$$\sigma = Ce^{\frac{-Et}{\eta}} \qquad (5\text{-}13)$$

In Equation 5-13, C is a constant of integration that can be determined by noting the boundary condition:

$$\sigma(t=0)=\sigma_0 \qquad (5\text{-}14)$$

that finally yields

$$\sigma = \sigma_0 e^{\frac{-Et}{\eta}} \qquad (5\text{-}15)$$

For functions that decay exponentially, another variable known as the time constant τ, is often introduced to characterize the rate of decay:

$$\sigma = \sigma_0 e^{\frac{-t}{\tau}} \qquad (5\text{-}16)$$

where

$$\tau = \frac{\eta}{E} \qquad (5\text{-}17)$$

While the Voight model suffers from an inability to predict any useful response on its own, it can be combined with other elements to form the basis of more accurate compound models such as the three-element model below.

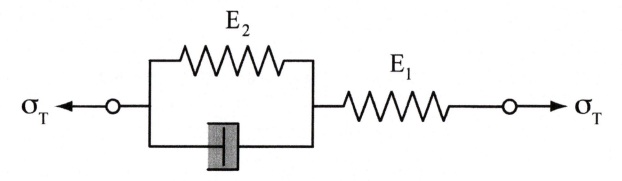

Figure 5-7. Three element model.

If "1" denotes spring #1, and "2" spring #2, then by equilibrium,

$$\sigma_T = \sigma_1$$

and

$$\sigma_T = \sigma_2 + \sigma_D \quad [\text{see Equation. } (5\text{-}5)] \qquad (5\text{-}18)$$

Deformation compatibility yields:

$$\varepsilon_T = \varepsilon_1 + \varepsilon_2 = \varepsilon_1 + \varepsilon_D \quad \text{since } \varepsilon_2 = \varepsilon_D \qquad (5\text{-}19)$$

The three constitutive equations relevant to the model are:

$$\sigma_1 = E_1 \varepsilon_1$$

$$\sigma_2 = E_2 \varepsilon_2 \qquad (5\text{-}20)$$

$$\sigma_D = \eta \frac{d\varepsilon_D}{dt}$$

80

Once again, substituting Equation (5-20) into Equation (5-19), with the appropriate differentiating, and using equilibrium conditions of Equation (5-18), one obtains the following differential constitutive equation for the three element model:

$$\frac{d\varepsilon}{dt} + \frac{E_2}{\eta}\varepsilon = \frac{1}{E_1}\frac{d\sigma}{dt} + \frac{\sigma}{\eta}\left[1 + \frac{E_2}{E_1}\right] \qquad (5\text{-}21)$$

For the structural/test application of stress relaxation, the boundary conditions apply:

$$\frac{d\varepsilon}{dt} \equiv 0$$

$$\sigma(t=0) = \sigma_0 \qquad (5\text{-}22)$$

$$\varepsilon = \varepsilon_0$$

and for instantaneously applied strain ε_O,

$$\sigma_0 = E_1\varepsilon_o \qquad (5\text{-}23)$$

The stress relaxation equation for the three element compound model is:

$$\sigma(t) = \varepsilon_o\left[E_1 + \frac{E_2^2}{E_1 + E_2}\left(1 - e^{-\frac{t(E_1 + E_2)}{\eta}}\right)\right] \qquad (5\text{-}24)$$

The fun adventure planned for you today (a.k.a, experiment) is a stress relaxation test of a well-known polymer, namely polypropylene that at room temperature (300 K) is greater than half of its melting temperature of 450 K; therefore, the viscoelastic response will be readily observable. Typical applications of polypropylene include pharmaceutical containers, auto battery cases, water ski tow ropes (it floats), luggage, and carpeting. Polypropylene has good chemical resistance at room temperatures, but is susceptible to degradation from ultraviolet rays unless stabilized. The Maxwell model will be applied to the stress relaxation behavior of polypropylene and the appropriateness of this approach (to polypropylene) will be determined.

PROCEDURE

- Measure the width, thickness, and gauge length of the polypropylene specimen. The gauge length can be estimated as the length between the fillets.

- Place the specimen in the grips of the testing machine. Set the deformation rate to 1"/min. Load the specimen to 250 lb (1110N). Record the time for this load to accumulate.

- Let the time at which the load has accumulated equal to zero. Record the force at 15 second intervals for 1 minute, then record the force at 30 second intervals until five minutes has elapsed.

- Unload the specimen and remove it from the testing machine.

ANALYSIS OF DATA

- Convert the load data in table 3 to stress data in table 4. Show the appropriate sample calculations.

- Plot stress versus time. Leave room to the negative side of t=0 to plot the time it took to bring the specimen to load.

- Determine the modulus of elasticity of polypropylene from the data in table 2.

- Calculate the viscosity, η and time constant, τ at t=60 seconds using equations 5-15 through 5-17.

- Using the previously calculated values of E and η, generate theoretical values of stress for t=0, 60, 120... seconds using Equation 5-15 or 5-16. Show a sample calculation.

- Place the theoretical values in table 5. Take the natural log as indicated. Transfer the appropriate experimental data from table 3. Take the natural log of the empirical data.

- Plot the theoretical data ($\ln\sigma_{THR}$ vs. time). The theoretical data should be a straight line. Superimpose the experimental data points.

- Calculate η at t=120 seconds.

1. Given below is a typical stress vs. time relaxation curve for ASTM A 278-51T cast iron at an elevated temperature. Assuming that this material, at this temperature, is accurately predicted by the Maxwell Model, determine the parameters σ_O, τ, and η. The Young's Modulus for steel varies very little with composition and can be assumed to be equal to 207 GPa.

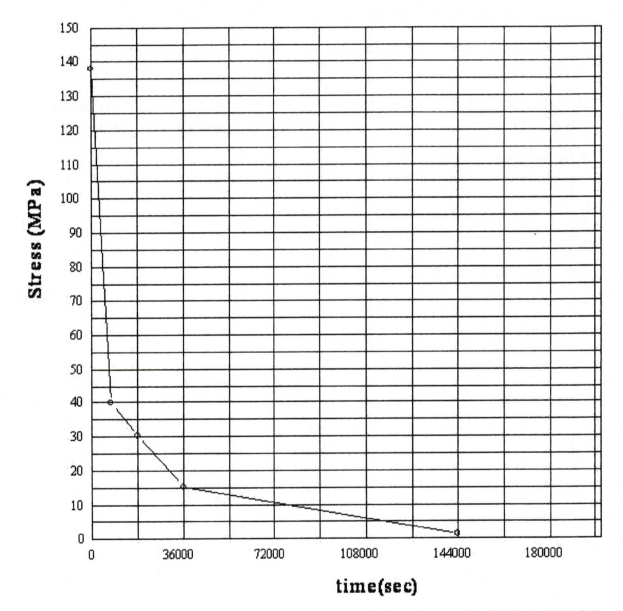

Figure 5-8.   Stress relaxation curve for ASTM A 278-51T cast iron at T=593°C (ASTM DS 60, 1982).

LAB #6.

STEADY STATE CREEP

INTRODUCTION

Inevitable jokes aside, Creep is defined as a material response where strains accumulate with time while under a constant load or stress and at a given temperature; the previous lab was an example of this phenomena at lower temperatures. Creep is clearly temperature dependent and is usually of sufficient concern to engineers when the operating temperature exceeds 30% of the material's absolute melting or glass transition temperature ($T > 0.3T_{MELT}$ or $T > 0.3T_{TRANS}$).

Most creep failures can be categorized into two main types: a tolerance failure called "strain-limited design" or a catastrophic failure usually called "creep rupture." As its name implies, a tolerance failure occurs when a part creeps (usually, but not exclusively) forward in length and exceeds a tight clearance. Classic examples are moving parts in engines and turbines. These parts usually spin in housings with small tolerances for increased efficiency; once they begin to creep and interfere with the housing, problems and failures will occur. Another example is the car door left open for a long period of time where the hinges creep and the door will no longer close properly. On the other hand, creep rupture is primarily found in high temperature industries such as chemical synthesis and power generation plants. For these applications, sagging pipelines may not be critical, but rupture of the pipeline, especially for toxic waste in the vicinity of operators will definitely ruin their day and is intolerable.

Assuming that creep is measurable, that is both the stress (tensile for extension and compressive for contraction) and temperature are sufficiently high, a typical strain-time curve under tensile stresses would be observed as illustrated in Figure 6-1.

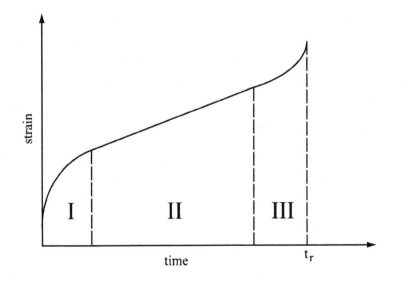

Figure 6-1. Typical creep curve with three response regions.

97

Figure 6-1 is divided into three distinct regions. Region I is defined as primary or transient creep that occurs at the start of the process. In this region, the strain rate initially starts out relatively high, but eventually decreases with time. Region II is called secondary of steady state creep; creep strains in this region (that accounts for most creep deformation) accumulate linearly with time. The final region, III, is tertiary creep and represents the onset of instability and rupture at time, t_r. Clearly, the creep process by its very definition is a time dependent phenomena. In fact, certain high temperature alloys can creep at very slow rates that may takes months or even longer to show any appreciable deformations. As such, you may want to avoid such topics in grad school or work unless you are ready for the "long haul."

Unfortunately, predicting creep can be an even more formidable task. Not unexpectedly, a large variety of predictive equations exist that can be found in many fracture mechanics texts. This experiment concentrates on steady state creep using one relatively simple predictive relationship, namely:

$$\dot{\varepsilon}_{ss} = B\sigma^m \qquad\qquad (6\text{-}1)$$

Equation 6-1 is known as the power law for creep prediction and can be used for quick estimations of steady state creep.

For the thrilling experiment to come, the creep rate of polypropylene will be examined. Four identical specimens will be dead weight loaded using different weights on a frame that features a simple lever for load magnification. Thus, a variety of stress magnitudes will be applied to the specimen. Dial gauge extensometers and a timing device will be used to measure elongation to determine the relationship of creep rate to stress amplitude. Finally, this data will be pooled and used to determine the material constants of the power law for creep prediction.

PROCEDURE

- Measure and record all relevant dimensions of the loading frame. Measure and record the width and thickness of the polypropylene specimen. Use the metric micrometer whenever possible.

- Place a specimen in the loading frame and use an allen wrench to tighten the grips.

- Mount the extensometer in a nearly fully compressed position on the specimen. Measure the gauge length.

NOTE: the following three steps are to be completed as near to simultaneous as possible. *Remember dropped weights break toes and will not do much to put (or keep) a smile on your face and a song in your heart. Hence, be very very careful!*

- One student should hold the arm of the loading frame up while another lab assistant (Igor, perhaps) applies the weights to the hanger. Release the weights smoothly so that there is a continuous transition from the unloaded to the loaded state. Remember, d**o not drop the weights** ☠

- Zero the extensometer.

- Record the extension as zero at time equal to zero.

- Take elongation readings every minute for fifteen minutes, then every five minutes until a total of sixty minutes has elapsed. Use your time wisely; begin the data analysis section.

ANALYSIS OF DATA

- Calculate the area of the specimen and the stress in the specimen.

- Convert the elongation to strain. Show a sample calculation. Place the values in table 3.

- Plot strain vs. time. Indicate the constant load and stress on your plot label or legend.

- Determine the steady state strain rate, $\dot{\varepsilon}_{ss}$ graphically.

- Pool your data on the chalkboard (stress and steady state strain rate). Complete table 4.

- Plot steady state strain rate vs. stress on log-log paper. Draw in a best fit line. Determine the material constants in equation 6-1. Remember (m) is the slope of the best-fit line and (B) should be calculated from a point that lies on the best fit line. Are your units correct?

1. Below is a table of strain rate and stress data during a steady state creep experiment. Plot strain rate versus stress on **log-log** paper. Determine the equation constants m and B from the best fit line.

Stress (MPa)	Strain rate (1/sec)
40	0.3×10^{-6}
60	0.5×10^{-6}
70	2×10^{-6}
150	7×10^{-6}

LAB #7.

PREDICTION OF YIELDING
FOR GENERAL STATES OF STRESS

INTRODUCTION

Yielding occurs when a material experiences permanent or "plastic" deformations that are not related to cracking or some other form of failure. Given the extensive usage of ductile alloys, predicting when a structural element, or an item being fabricated yields is one of the most basic of engineering analyses. Many if not most items such as precision machinery must remain elastic while being used. Conversely, forming paper clips out of pieces of wire is most competently done by actually plastically straining the wire, lest you end up with the same straight piece of wire (and probably a pink slip from your employer). In both cases, the ability to determine the transition from elastic to plastic is essential.

You may fondly recall that in Lab #3 (tensile behavior of a ductile material) provided the data necessary to determine the 0.2% offset yield strength of a ductile material when loaded in tension; usually the same numerical result holds in pure compression, or one simply assumes that it does. Yield strength is not an inherent property of a ductile material as experience has clearly shown that it depends on the nature of the stress state applied. Thus, for other modes of loading not as simple as pure tension, there are two options for determining the onset of yielding. The first and perhaps most obvious is to characterize yielding experimentally. However, this would require recreating the exact loading and stresses for the component (and material) in question and then tediously looking for permanent strains; indeed (and mercifully), this was the reason that the 0.2% offset method was originally developed to determine yield strength. Another option is to predict yielding via an appropriate and hopefully accurate theory. In reality, this is the only economically viable approach.

On the atomistic level, yielding is known to be due to the motion of dislocations and crystalline line imperfections as driven by shear stresses acting on their slip planes and in the direction of their Burger's vector. Other factors being constant, the critical resolved shear stress is the property of a single crystal that describes yield strength. However, for polycrystalline metals, a parameter that is dependent on the total stress state would be more appropriate. The load parameters can be amply described by knowing the three independent principal stresses at points of concern in the structure, notably where they are at their "*worst*," whatever that means.

There are two main theories to predict yielding that are both based on shear considerations. The first is generally attributed to Tresca, and is often referred to as the Maximum Shear Stress Theory (MSST). Simply put, MSST states that yielding occurs when the maximum shear stress at any point equals a critical value, K, or:

$$\sigma_I - \sigma_{II} = \pm K$$
$$\sigma_{II} - \sigma_{III} = \pm K \qquad\qquad (7\text{-}1)$$
$$\sigma_{III} - \sigma_I = \pm K$$

Whichever of these three equations has the largest term on the left is the appropriate one. If a tension test is used to evaluate the factor K, then one may set two of the principals equal to zero - e.g., $\sigma_I = \sigma_{II} = 0$-and the other, σ_{III}, equal to the offset yield strength σ_y. Thus:

$$\sigma_I - \sigma_{II} = \pm\sigma_y$$
$$\sigma_{II} - \sigma_{III} = \pm\sigma_y \qquad (7\text{-}2)$$
$$\sigma_I - \sigma_{III} = \pm\sigma_y$$

However (and this is a potential life or time saver), if the principal stresses are designated such that $\sigma_I > \sigma_{II} > \sigma_{III}$, then one only has to look at:

$$\sigma_I - \sigma_{III} = \pm\sigma_y \qquad (7\text{-}2a)$$

since it will always represent the largest difference or more importantly, the maximum shear stress! Note that any intermediate shear stresses on the other planes do not come into play.

An alternative approach attributed to von Mises, but also known as the Distortion Energy Theory (DET) or the Octahedral Shear Stress Theory, uses a higher order function of the maximum shear stresses, or differences in principal values, and can be written:

$$(\sigma_I - \sigma_{II})^2 + (\sigma_{II} - \sigma_{III})^2 + (\sigma_{III} - \sigma_I)^2 = K \qquad (7\text{-}3)$$

If K is evaluated in tension as above, $K = 2\sigma_y^2$, and:

$$(\sigma_I - \sigma_{II})^2 + (\sigma_{II} - \sigma_{III})^2 + (\sigma_{III} - \sigma_I)^2 = 2\sigma_y^2 \qquad (7\text{-}4)$$

A comparison of the two criteria just discussed (MSST and DET) are shown in Figure 7-1 in a planar, principal stress-space (σ_{III} was arbitrarily selected to be zero). Every stress state within the figure(s) is elastic, subject to yielding on the boundary, and elastic-plastic outside. Note that although they are in agreement at six points, the von Mises criteria allows larger stresses before yielding than the Tresca. Hence, MSST can be considered conservative from a design standpoint. Furthermore, the only material parameter necessary to exploit either one of the theories is the offset yield strength measured in a very simple tension test.

This lab will employ a multi-axial stress test that will be used to compare the accuracy of the two yield theories just discussed. As shown in Figure 7-2, the loading mode will be pure shear; an aluminum alloy (Al 6061-T6) shaft with a circular cross section will be loaded in torsion. In order to keep the shear stress invariant across the entire cross-section, a thin walled tube will be used during the test. For this configuration, the formula for shear stress in a thin-walled bar undergoing a torsional load where, $\bar{r} = r_o - t/2$ is the bar's mean radius:

$$\tau_{xy} = \frac{T}{2\pi\bar{r}^2 t} \qquad (7\text{-}5)$$

116

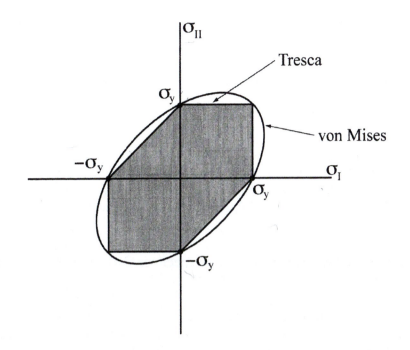

Figure 7-1. Tresca and von Mises yield envelopes for plane stress.

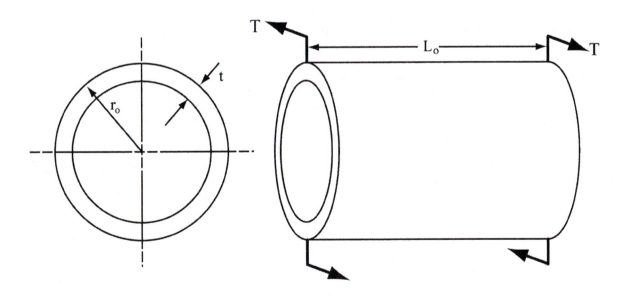

Figure 7-2. Loading schematic of thin-walled tube used to assess yielding.

117

PROCEDURE

• Measure and record the outer diameter and wall thickness of the aluminum tube.

• Insert an end plug to each end of the tube; failure to do this will ruin the specimen and perhaps your day if the instructor finds out ☠. Place the specimen with plugs into the torsion machine. Be sure to use the full length of the grips. Tighten the grips until they engage the specimen. The grips need not be tightened extensively, for the grips are self tightening upon loading.

• Measure and record the gauge length of the specimen defined in this situation as the distance between the machine's grips.

• Rotate the hand crank until a small initial torque is indicated on the gauge. The corresponding angular displacement will be set equal to zero. All further angular displacements will be taken with respect to this zero.

• Continue twisting the specimen by 4 degree increments until the relative angular displacement equals 40 degrees. Record the angle and torque values. NOTE: Further displacement will excessively yield the tube. The tube will then buckle and the end plugs will jam; *groups doing this will responsible for cutting out the end plugs and all of the joys (and expenses) that come with the task.*

• Unload the specimen and remove it from the machine. Discard the specimen appropriately.

ANALYSIS OF DATA

- Plot torque versus angle of twist.

- Determine the offset angle of twist corresponding to the 0.2% offset shear strain from:

$$r_o \theta = \gamma L_0 \qquad\qquad (7\text{-}6)$$

  where θ is in radians.

- Draw the 0.2% line and graphically determine the offset yield torque. Convert this torque to a shear yield strength; use equation 7-5.

- Redraw the MSST and DET yield envelopes. Be sure they are properly scaled. Also indicate the two possible loading paths. Tabulate the two theoretical and one experimental measures of yield strength in shear determined in this lab.

1. Below is a table representing a portion of typical stress-strain data for an Al 6061-T6 aluminum tensile specimen. Plot the data and determine the modulus of elasticity and the 0.2% offset yield strength.

Table 1. Typical stress-strain data

Stress (MPA)	Strain (m/m)	Stress (MPA)	Strain (m/m)
0	0	220	0.00360
50	0.00072	235	0.00405
100	0.00145	245	0.00443
150	0.00217	255	0.00505
175	0.00254	270	0.00599
200	0.00306	275	0.00740

2. Construct the plane-stress yield envelopes in a principle stress space for both the Tresca and the von Mises yield theories using your calculated value of the yield strength to scale the envelopes. Indicate the two equivalent load paths corresponding to pure shear on the yield envelopes. Calculate the shear yield strength of Al 6061-T6 aluminum predicted by the above theories.

Appendix A.

MECHANICS OF EXPERIMENTAL
STRESS AND STRAIN ANALYSIS

Whether they are obtained analytically, experimentally, or numerically, the components of stress, σ or strain, ε at a point consist of nine components. Six of these components are independent and can be displayed in 3 x 3 arrays as:

$$\left|\sigma_{ij}\right| = \begin{vmatrix} \sigma_{xx} & \tau_{xy} & \tau_{xz} \\ \tau_{yx} & \sigma_{yy} & \tau_{yz} \\ \tau_{zx} & \tau_{zy} & \sigma_{zz} \end{vmatrix} \qquad \left|\varepsilon_{ij}\right| = \begin{vmatrix} \varepsilon_{xx} & \dfrac{\gamma_{xy}}{2} & \dfrac{\gamma_{xz}}{2} \\ \dfrac{\gamma_{yx}}{2} & \varepsilon_{yy} & \dfrac{\gamma_{yz}}{2} \\ \dfrac{\gamma_{zx}}{2} & \dfrac{\gamma_{zy}}{2} & \varepsilon_{zz} \end{vmatrix}$$

The stresses are shown on the element below even though the element actually represents a *point* in a structural element. By moment equilibrium: $\tau_{xy} = \tau_{yx}$, $\tau_{xz} = \tau_{zx}$, and $\tau_{yz} = \tau_{zy}$. Similarly, the shear-strain definitions ensure that $\gamma_{xy} = \gamma_{yx}$, $\gamma_{xz} = \gamma_{zx}$, and $\gamma_{yz} = \gamma_{zy}$.

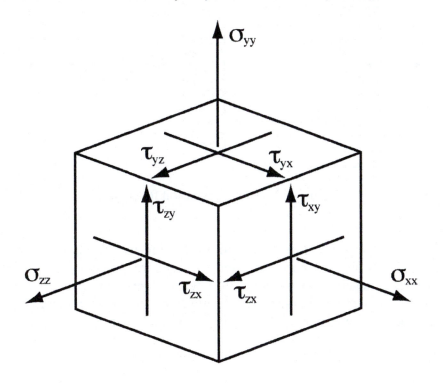

Figure A-1. Three-dimensional state of stress at a point.

Solving for, or measuring stress or strain components is usually done with the geometry of the structure and the directionality of the loading fixing the coordinate system. For example, if the element in question is a beam, one axis is usually selected to coincide with the beam axis with the other in the loading direction; the third direction is then fixed by the right-hand rule. However, the stresses or strains one obtains with these *natural* axes are not commonly the extreme values used in design. Instead, having found the state of stress or strain at that point, one rotates the axes about their origin, searching for axis orientations that correspond to either the maximum shear stress or the maximum normal stresses.

STRESS TRANSFORMATIONS

The extreme values of stress, and the axis orientations at which they act may be found from the appropriate stress equilibrium equations; these equations are commonly known as the *Stress Transformation* Equations. However, before the transformations are done using the much-loved graphical device known as Mohr's Circle, the appearance of the transformation equations in a normal/shear stress space, it is first necessary to establish sign conventions for stress.

Since stresses involve both forces and planes, there will be two contributions to their signs. Consider the stress element with the two planes shaded as shown below in Figure A-2:

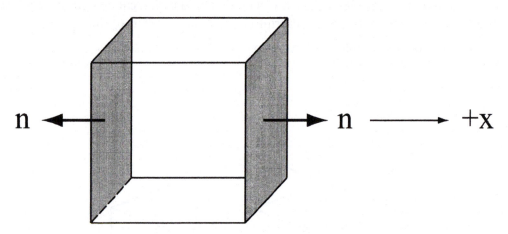

Figure A-2. Sign convention for planes.

The normal vector to the plane, **n**, is drawn outwardly-directed; its sign depends on whether it points in the positive or negative coordinate sense. The normal vector is, therefore, positive on the right plane while negative on the left. The force component also has a vectorial sense. Hence, the sign of any stress component is determined by (a) the sign of the plane's outwardly directed normal and (b), the sign of the force component such that:

Stresses are *positive* when (a) and (b) above are the same sign.

Stresses are *negative* when (a) and (b) above have different signs.

In order to use Mohr's circle to search out extreme stress values, it is first necessary to define the *twisting sense* of the shear stresses. For example, in the case of plane stress, there are two physically distinct possibilities for the stresses: they are either positive or negative as shown in figure A-3.

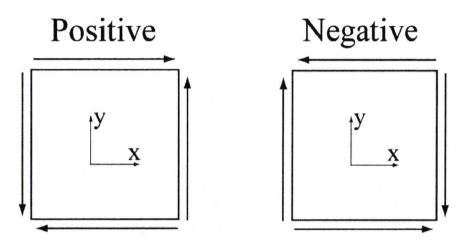

Figure A-3. Positive and negative states of pure planar shear stress.

Consider the positive example that can be decomposed into the stresses on the x- and y-planes as shown below:

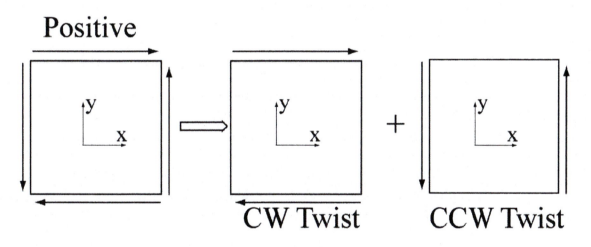

Figure A-4. The twisting sense of a state of positive pure shear stress.

Note that the stresses are identified by their *twisting sense* exerted on the stress element, with τ_{xy} having the opposite sense (always) of τ_{yx}. Recall that the shears are still algebraically equal.

A *Mohr's Circle* for stress is constructed below for a state of plane stress: the example depicted is for the case $\sigma_{xx} > \sigma_{yy} > 0$; $\sigma_{zz} = 0$; $\tau_{xy} > 0$.

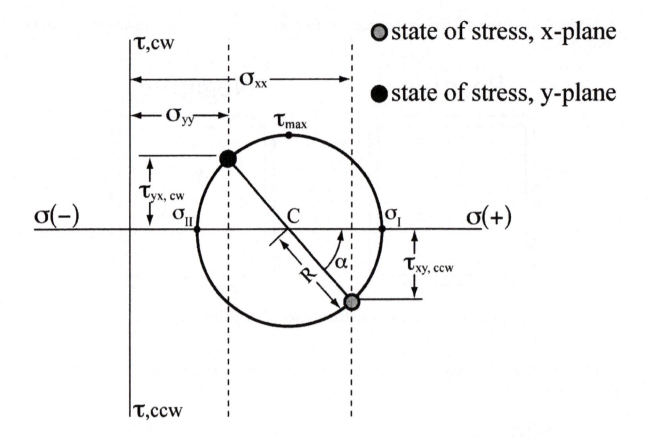

Figure A-5. Mohr's circle for plane stress transformations.

A circle is constructed by using the normal and shear stresses for the two planes ar x-y pairs and plotting them accordingly. Hence, the pair (σ_{xx}, τ_{xy}) defines the x-plane while (σ_{xx}, τ_{xy}) represents the y-plane. Obviously, the stresses could also involve the x-z or y-z planes.

The extreme values of stress are the max/min normal stresses σ_I and σ_{II} and are commonly known as *Principal Values* or *Principal Stresses*. In addition, the maximum shear stress τ_{max} is the load variable of interest in certain design situations. If C is the center of the circle and R is its radius, then:

$$\sigma_I = C + R$$

$$\sigma_{II} = C - R$$

$$\tau_{max} = R$$

wherein:

$$C = \frac{\sigma_{xx} + \sigma_{yy}}{2}; \qquad R = \sqrt{\left(\frac{\sigma_{xx} - \sigma_{yy}}{2}\right)^2 + \tau_{xy}^2}$$

Having established the extreme value stresses numerically, one then has to specify the planes upon which they act. To do this, it is necessary to recall that rotations in the real space coordinate system, clockwise or counterclockwise about the z-axis, are done in the same rotational sense on Mohr's circle, except real angles are doubled on the circle. In the above example, the reader can see that to get from the point representing the stress state on the x-plane to the principle state featuring σ_I, you must rotate counterclockwise by α degrees; in real space, the rotation will be half, but in the same sense as illustrated by Figure A-6

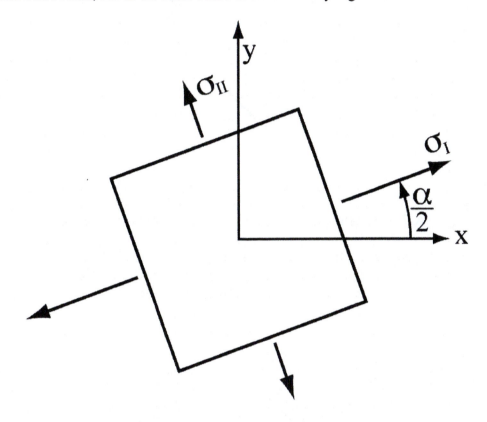

Figure A-6. Principal stress element located relative to the x,y-axes

The state of maximum shear stress, as well as the planes upon which it acts, is located at the top/bottom of the circle, always 90° away from the principal planes. If one rotates 90° counterclockwise from the planes whose normal vectors are in the I-sense, the maximum shear stress has a CW twist sense, and acts as shown in Figure A-7

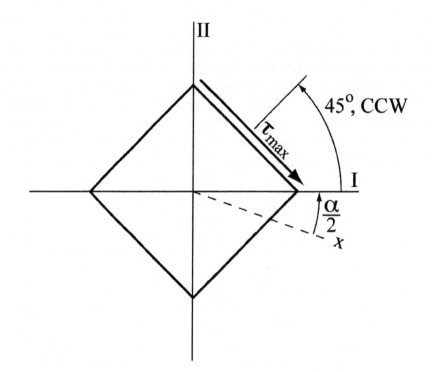

Figure A-7. Twisting sense and plane of action of Maximum Shear Stress

The stress element with the complete state of maximum shear stress is shown below. Note that there is a normal stress component for the point on the top (bottom) of the stress circle that is always equal to "C" of the circle defining the maximum shear stress.

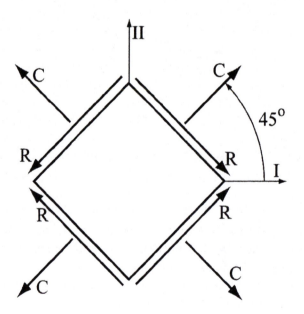

Figure A-8. Complete Maximum Shear Stress State.

ELASTIC CONSTITUTIVE EQUATIONS

Strains as defined as changes in line length or angle, are often the quantities experimentally determined in experimental mechanics. Stresses are then readily calculated from the strain measurements using constitutive relationships. In this case, a number of important assumptions must be made about the way in which one expects the material to act. First, it will be assumed that the material is *isotropic* so that any relevant materials parameters do not depend upon direction. Materials that are not isotropic are called *anisotropic* and can still be analyzed, but the algebra is far more complex (aka, a royal pain). Fortunately, isotropy is a useful and reasonable approximation for many materials, particularly metals. Secondly, it will be assumed that the material responds linearly, or that forces are linearly related to deformations as in the spring equation $F = kx$. Such mechanical response was first formalized in 1678 by Robert Hooke:

"Ut tensio, sic vis"

Unfortunately for the engineers of the time, it was a few hundred years before "all the stretches are proportional to all the pulls, and vice versa" was put into the now familiar algebraic form.

Recall that the axial stress and strain are related in a tension test by:

$$\sigma_{ax} = E\varepsilon_{ax}$$

where E is the Young's modulus, or, alternatively, the Modulus of Elasticity, and that there is a transverse strain in a tension test related to the axial strain by:

$$\varepsilon_{trans} = -\nu\varepsilon_{ax} = -\nu\frac{\sigma_{ax}}{E}$$

wherein ν is the Poisson ratio. The fully three-dimensional form of Hooke's law for stresses and strains can now be assembled as:

$$\varepsilon_{xx} = \frac{\sigma_{xx}}{E} - \nu\frac{\sigma_{yy}}{E} - \nu\frac{\sigma_{zz}}{E} = \frac{1}{E}\left[\sigma_{xx} - \nu\left(\sigma_{yy} + \sigma_{zz}\right)\right]$$

$$\varepsilon_{yy} = \frac{1}{E}\left[\sigma_{yy} - \nu\left(\sigma_{zz} + \sigma_{xx}\right)\right]$$

$$\varepsilon_{zz} = \frac{1}{E}\left[\sigma_{zz} - \nu\left(\sigma_{xx} + \sigma_{yy}\right)\right]$$

$$\gamma_{xy} = \frac{1}{G}\tau_{xy} \quad \gamma_{yz} = \frac{1}{G}\tau_{yz} \quad \gamma_{zx} = \frac{1}{G}\tau_{zx}$$

wherein G is the shear modulus or, alternatively, the Modulus of Rigidity. For isotropic solids, G can be related to E and ν by the following relationship:

$$G = \frac{E}{2(1+\nu)}$$

If strains are being measured on a free surface by conventional means (strain gauges for example), then the normal and shear stresses having a subscript in the same sense as the perpendicular to that surface are zero, thus simplifying the equations. In addition, since the stresses are being calculated from measured strains, it will be convenient to simultaneously solve the above equations once and for all.

Since strain gauges are generally on the surface of a structure, a plane strain condition exists ($\varepsilon_{zz}=0$). Letting the plane of measurement be the x,y-plane, and the measured strain quantities then being ε_{xx}, ε_{yy}, and γ_{xy}, the stresses are computed by:

$$\sigma_{yy} = \frac{E}{(1-v^2)}\left(\varepsilon_{yy} + v\varepsilon_{xx}\right)$$

$$\tau_{xy} = G\gamma_{xy}$$

Appendix B.

REFERENCES

Budinski, K, Engineering Materials Properties & Selection, Prentice Hall, Englewood Cliffs, NJ, 1989.

Caddell, Robert, Deformation and Fracture of Solids, Prentice Hall, Englewood Cliffs, NJ, 1980.

Queeney and Segall, Mechanical Response of Engineering Materials, 2nd Edition, The Pennsylvania State University, University Park, PA, 2009.

Compilation of Stress-Relaxation Data for Engineering Alloys, ASTM Data Series Publication DS60, ASTM, Philadelphia, PA, 1982, p 15.

Dally and Riley, Experimental Stress Analysis, Mc Graw-Hill, New York, 1991.

DiMassa and Queeney, Experiments in the Determination of Mechanical Behavior of Engineering Materials, third edition, The Pennsylvania State University, University Park, PA, 1989.

Halliday and Resnick, Physics, John Wiley and Sons, New York, NY, 1978.

Hertzberg, Deformation and Fracture Mechanics of Engineering Materials, John Wiley and Sons, New York, NY, 1989.

Higdon et. al., Mechanics of Materials, John Wiley and Sons, New York, NY, 1985.

Metals Handbook Ninth Edition Properties and Selection: Nonferrous Alloys and Pure Metals, ASM, Metals Park, OH, 1979.

Zandman, Redner and Dally, Photoelastic Coatings, Iowa State University Press, Ames, Iowa, 1977.

CPSIA information can be obtained at www.ICGtesting.com
Printed in the USA
LVOW02s0811150815

449736LV00006B/12/P